PLANET IN PERIL

FIERCE FLOODS

Cath Senker

First published in 2017 by Wayland
Copyright © Hodder and Stoughton 2017

Wayland, an imrprint of Hachette Children's Group
Part of Hodder & Stought
Carmelite House
50 Victoria Embankment
London EC4Y 0DZ

Editor: Elizabeth Brent
Designer: Ray Bryant
Cover design by Rocket Design (East Anglia) Ltd.

Dewey number: 363.3'493-dc23
ISBN   978 0 7502 8911 5
eBook ISBN   978 0 7502 8547 6

Printed in China

10 9 8 7 6 5 4 3 2 1

Picture acknowledgements: p5 (t) © smallroomphoto/Shutterstock.com; p6 © Getty Images (b); p8 © Jashim Salam / Demotix/Demotix/Corbis; p9 (t) © 2009 Jonas Gratzer, (b) © AFP/Getty Images; p10 © ANDREW BIRAJ/Reuters/Corbis; p11 (l) © ANDREW BIRAJ/Reuters/Corbis; p12 © MUNIR UZ ZAMAN; p13 (t) © UIG via Getty Images; p15 © Getty Images; p16 © SEAN GARDNER/Reuters/Corbis; p17 (b) © Barbara Gauntt/ZUMA Press/Corbis; p18 © HO/Reuters/Corbis; p19 (t) © Getty Images; p20 © AFP/Getty Images; p21 (t) © NOAA, (b) © Matytsin Valery/ITAR-TASS Photo/Corbis; p22 © Orlov Grigory/ITAR-TASS Photo/Corbis; p23 (t) © RIA Novosti/Reuters/Corbis, (l) © Palm Yelena/ITAR-TASS Photo/Corbis, (r) © Orlov Grigory/ITAR-TASS Photo/Corbis; p24 © Chumash Ivan/ITAR-TASS Photo/Corbis; p25 (t) © AFP/Getty Images, (b) © Chumash Ivan/ITAR-TASS Photo/Corbis; p27 (l) © Brooks Kraft/Sygma/Corbis, (r) © GASTON DE CARDENAS/Reuters/ Corbis; p28 © National Park Service; p29 (l) © ibrahim/Demotix/Corbis. All other images, including the cover image, courtesy of Shutterstock.com.

Text acknowledgements: p11 Eyewitness: 'Families struggle to rebuild homes and livelihoods following floods', Joe Cropp and Maherin Ahmed, International Federation of Red Cross and Red Crescent Societies, 4 October 2012; p13 Case study: Christian Aid: Superducks; p17 Eyewitness: 'Mississippi flood evacuees spend tedious days in shelter, Times-Picayune Staff, The Times-Picayune Greater New Orleans, 18 May 2011; p23 Eyewitness: 'About 280 millimeters of rain fell in Gelendzhik', Encyclopedia of Safety; p25 'Russia convicts officials of 2012 flood negligence', Agence France-Presse, 21 August 2013; p27 Expert view: 'Could the "biblical" northern Colo. floods have been predicted?' Stephanie Paige Ogburn, E&E Publishing, 26 September 2013; p29 Expert view: 'Canada slow to initiate disaster prevention programs, experts warn', Jessica Barrett, Calgary Herald, 23 September 2013.

Wayland is a division of Hachette Children's Group, an Hachette UK company.
www.hachette.co.uk

Contents

What are floods?   4

Accidental and deliberate flooding   6

Bangladesh, June 2012   8

The Mississippi River, May 2011   16

Russia, July 2012   20

Prediction and protection   26

The storm-water challenge   28

Glossary   30

Find out more   31

Index   32

What are floods?

Have you ever been in a huge storm with pouring rain? The rain falls in rivers, soaks into the ground or flows down the drains. Floods happen when the ground or drains cannot soak up all the rain or rivers cannot cope with the extra water. Floods sometimes take a long time to develop, allowing people time to prepare or evacuate. At other times, they arrive quickly and catch people by surprise. River floods are the most common kind of flood. They occur when water bursts over riverbanks. Flash floods are sudden downpours of rain in a small area. Coastal floods happen where the land at the coastline is just a few metres above ground, and the sea floods over it.

RIVER FLOODS

Intense rainfall over a short space of time can cause river floods. Sometimes, ice melts rapidly in nearby mountains, and vast quantities of water surge into the river. The water builds up as it moves down the river. If the water rises higher than the riverbanks, it spills over and covers the floodplain — the land surrounding the river. River floods are often seasonal. For example, in South Asia, the summer monsoon rains regularly lead to flooding.

COASTAL FLOODS

Along coastlines, storm surges are the major reason for flooding. Huge storms with mighty winds, such as tropical cyclones, cause the sea to rush inland. High waves batter the coast, and may flow up estuaries, creating floods inland too. Low-lying coastal regions, such as the Maldives in the Indian Ocean, are at greatest risk.

FLASH FLOODS

Heavy rainfall from a thunderstorm or cloudburst — a sudden, violent rainstorm — can create flash floods. Flash floods often occur on short rivers that cannot absorb a large amount of rain. The water crashes over the riverbanks and seeks the quickest route downhill; this might be through a town or a narrow valley. Low-lying coastal areas, for example, in Bangladesh, are also prone to flash flooding.

Accidental and deliberate flooding

HELPFUL FLOODING

When the floodwaters drain away, they leave the area blanketed in sediment — mud and silt from the river. This sediment is usually full of nutrients that improve the fertility of the soil, making it good for growing crops. For example, the floodplain of the River Nile in Egypt is flooded every year, and the fertile land alongside the river has been used for agriculture for thousands of years. The Mississippi River in the mid-west of the USA plays the same role.

FLOOD HAZARDS

While floods can be useful, their devastating effects are often greater. The force of the water destroys bridges and buildings, uproots trees, drags away vehicles and drowns people caught in its path. The floodwaters carry poisonous chemicals, fuel, sewage and debris, which contaminate the landscape. The stagnant (still) water rots buildings and bacteria breed in it, which can cause dangerous diseases. Floods kill more people than any other natural disaster and cause as much damage to property as all other kinds put together.

CLIMATE CHANGE AND FLOODS

Nowadays, we see greater numbers of severe floods than ever before. Scientists link this to climate change. The world is becoming warmer, and hotter weather makes tropical cyclones larger, with stronger winds and heavier rainfall. They cause more frequent and violent storm surges over wider areas. The melting of the ice caps is making sea levels rise, so coastal regions run higher risks of flooding. This book looks at dramatic examples of coastal, river and flash floods, their effects and how communities recovered from them. It considers how technology can help us to predict floods and reduce their devastating impact.

FACT BOX — How tides affect flooding

The pull of gravity from the Moon creates tides on Earth. The Moon pulls on the water in the oceans, causing it to bulge. As the Moon orbits (circles) our planet and the Earth rotates, the bulge moves and where the water bulges, it is high tide. The Sun's gravity also pulls on water, but more weakly because it is so far away. When the Sun and the Moon both pull on the ocean in the same place at the same time, high tide is even higher than normal. At this time, any floods will be more powerful because the water is already above the usual level.

The people of Bangladesh are used to the monsoon floods — intense rains over a short period of time that come every summer, leaving fertile sediment behind. However, no one had seen anything like the monsoon that hit in late June 2012. It was the heaviest rainfall in many years, and caused severe flash flooding on the coast. Chittagong, the country's second largest city, saw 463 mm (18 in) of rain in 24 hours — close to the average for the entire month! Torrential rain continued for three days.

The people of Chittagong could not go to work. It was almost impossible to walk around because the water was so deep. Bridges collapsed, and trains could not run in and out of the city. Mudslides buried houses with people still inside them.

WHY ARE THE FLOODS SO BAD?

Two-thirds of Bangladesh lies less than 5 m (16 feet) above sea level so it is prone to flooding. But over recent decades, human action has made the floods worse. In Nepal, deforestation — cutting down trees for timber, farmland and firewood — means there are fewer trees to soak up rainfall and prevent flooding. And climate change means that the floods in Bangladesh are now more violent and frequent as well as less predictable.

IMMEDIATE IMPACT

The 2012 floods had a dramatic impact. They arrived so suddenly that people had no time to prepare by storing food and household goods, and evacuating their homes. In Sylhet, in north-east Bangladesh, houses filled with up to 1 m (3 feet) of water. Many homes were made of mud and hundreds were simply washed away. Residents took to boats or rushed to higher ground. Farmland was flooded, and transport links ground to a halt.

Rainfall?
463 mm (18 in) of rain in 24 hours

How many killed?
139

How many homes damaged?
360,862

How many people displaced?
50,778

Crop area damaged?
93,248 hectares (230,421 acres)

Immediate aftermath

The floods and storms caused horrifying deaths from drowning, landslides, collapsing walls and direct lightning strikes. In Sylhet, 50,000 people were completely cut off by high waters after scrambling to high ground. Others trudged through waist-high water to relief camps or simply sheltered at roadsides.

The situation was desperate. With crops destroyed, fish swept away and other food supplies ruined, people went hungry. Wells and latrines (pit toilets) were damaged. People were forced to drink riverwater and go to the toilet in the open air, increasing the risk of disease.

A flood victim cooks a meal in front of her makeshift tent in a flooded village in Gaibandha.

Eyewitness

No home, no food, no water

Nomita Rani, a 40-year-old mother of four, used to live in a two-bedroom house in northern Bangladesh. It was completely washed away by the June 2012 floods. Three months later, she and her husband and their children were surviving by the roadside. Sometimes there was no fresh water, and they were forced to drink the dirty water around them, although they knew the risks. When Nomita ventured out to rescue possessions from her house or to find food, she found herself walking in water up to her neck.

She explained, 'We need a dry place to sleep, clean water to drink, food for our children. Sometimes we don't have anything to eat for days.' Nomita was visited by an assessment team, and they gave her a document saying she could collect a tarpaulin, clothes and some cash from the Bangladesh Red Crescent Relief Society in Shariakandi. It took her almost a day to walk there but she desperately needed the help.

Flood victims collect aid in Sylhet, north-east Bangladesh.

RUSH TO THE RESCUE

The Bangladeshi authorities responded fast, distributing food, drinking water, clothing, tarpaulin sheets for shelter, medicines and money to survivors. The army, emergency services, local officials and volunteers joined the rescue efforts. Serajul Huq Khan, the government's chief administrative official for Chittagong Division, explained: 'Our main challenges are now to find the dead bodies, give treatment to the wounded, and provide drinking water and food to the affected people.' Emergency workers searched for people trapped in their homes or buried under mud, often with nothing but their hands and basic tools. They often found dead bodies.

Long-term effects

Many survivors lost everything. Majeda Begum, a mother from Baze Telkupin island, on the banks of the Brahmaputra River, recounted, 'The water came up to my neck. My clothes and the little rice I had were all washed away.' Large numbers of people lost their crops and vegetable gardens in the floods, so they had to sell their cows and goats to buy food. With no assets, they were forced to become labourers. So many people were looking for work that wages went down. Farmers could get away with reducing the pay because they knew people had no choice but to accept. To make things worse, rice was scarce after the floods, and the price of this vital food rose by 30 per cent.

A HELPING HAND

Aid agencies stepped in to assist people to rebuild their homes and find new ways to make money, to avoid falling into poverty and debt. For example, local non-governmental organization Gana Unnayan Kendra, along with international charities Oxfam and Christian Aid, helped people on Baze Telkupin island to raise the bases of their houses to reduce flood damage. It provided people with cows so they could sell the milk. Oxfam gave money to the poorest families so that they did not have to sell their assets or take out loans to survive and could start farming again once the floodwaters ebbed.

A small donation from a charity may allow people to keep their land.

Case study

Duck farming

Noor Islam was a labourer from Sylhet but now he's a successful farmer with his own home and shop. The secret of his success is duck farming. Ducks love the wet climate and can survive flooding. Mr Islam received a loan from Christian Aid and partner organization Friends and Village Development. Now he sells between 200 and 400 eggs a day. Not only do the ducks provide eggs but they also eat harmful insects in the rice fields so Mr Islam doesn't need insecticides. They leave their droppings so he doesn't have to buy fertilizer. Mr Islam is making a good living and adapting to climate change at the same time.

Several charities are helping Bangladeshi farmers to run successful duck farms.

The autumn of 2010 in Iowa, Minnesota and Wisconsin was particularly wet. Over winter, heavy snow covered large sections of the Mississippi River basin, forming a mass of frozen water. The spring thaw began in February, and intense rainfall added to the rapidly melting snow. By mid-April there was already some flooding. Forecasters realized that a record flood was on the way. Flood-fight leader Major General John Peabody told all senior officials that flood-duty missions were top priority. In May 2011, the mass of water caused flooding on a scale not seen since the huge floods of 1937. Water surged over the banks of the Mississippi, flooding thousands of square kilometres of farmland and communities in the Mississippi River valley.

MASS EVACUATION

After receiving flood warnings, more than 1,000 people in north Memphis, Tennessee, evacuated their homes. On 9 May, the waters reached halfway up the houses in flooded neighbourhoods. It was expected that the flood would move downstream towards New Orleans. Fearing the levees might be breached in Baton Rouge and New Orleans, Louisiana, the authorities ordered thousands of people to evacuate from those cities.

During the floods, water overflowed the levees in Missouri, Arkansas, Mississippi and Tennessee. Although thousands heeded the warnings and fled, several people drowned in flash floods and the flooding of river tributaries (minor rivers) in Arkansas.

DISASTER AVERTED

However, the flooding was not as disastrous as it could have been. Having learnt from flooding in the past, the authorities had built a system of levees and spillways to contain and channel the water. (Levees are embankments – banks constructed to prevent rivers overflowing. Spillways are structures that channel the flow of water from levees so the water does not destroy them.) Good prediction technology allowed experts to work out when and where the water would breach its banks so they knew when to open up the channels to release the floodwater.

Rainfall?
38–51 cm (15–20 in)

How many killed?
24

How many homes damaged?
more than 21,000

How many people displaced?
more than 600,000

Cost to the economy?
US $3.4 billion

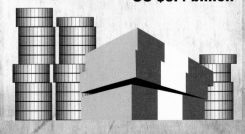

First response

The authorities acted fast, opening spillways to divert the water from large settlements. Quick thinking protected the town of Cairo, Illinois, where 2,660 people lived. On 2 May, engineers blew up part of a levee in Missouri to divert the floodwater away from Cairo. Instead, about 520 sq km (200 square miles) of farmland, inhabited by 200 people, were flooded. On 9 May, the Bonnet Carre Spillway was opened to allow the water to flow into Lake Pontchartrain, which drains into the Gulf of Mexico. Water was also released from Morganza Spillway, near Baton Rouge, allowing floodwater to drain into the Atchafalaya River basin.

DELAYS AND CLOSURES

These actions were successful, but the floods did disrupt economic life. The main grain-shipping port of Natchez, Mississippi, was closed for a few days. When it finally reopened, barges hauling valuable coal, timber, iron, steel and grain exports along the Mississippi had to travel as slowly as possible so that their wakes (waves created by movement) didn't increase strain on the levees. The exporters lost money because of the delays. Businesses were forced to close for a short time. For example, in Tunica, Mississippi, a big gaming centre, 17 out of 19 casinos closed. Yet the impact was not as bad as it could have been because of the effective flood-control measures.

 Eyewitness

Life on hold

Some of the worst flooding was in Vicksburg, Mississippi - 2,000 of the 4,800 people displaced in Mississippi were Vicksburg residents. They had to spend a couple of weeks in shelters. Vivian Taylor, a 60-year-old teacher, described her experience, 'When we saw water starting to build up in fields behind the neighborhood we started to get worried. Then we started seeing snakes and worms coming up out of the ground and we became very concerned.' Taylor spent the time exchanging stories with others at the church shelter. Apart from that, she recounted, 'I pray. I read. I meditate. I just try to sit calm and get my bearings.' There wasn't much else to do; the evacuees' lives were on hold.

Emergency workers look after a baby in a Red Cross Shelter for flood victims in Vicksburg, Mississippi.

Rebuilding and regenerating

After the flood, some homes had to be knocked down completely, or gutted (all the inside parts removed) and rebuilt. The water left debris that had to be cleared. Farming was seriously affected. Although the sediment boosted the nutrients in the soil, it also deposited debris, dangerous chemicals and bacteria. Crops on flooded land could not be harvested or new seeds immediately sown.

Geology professor Sam Bentley from Louisiana State University, Baton Rouge, noted: 'If you have a pasture where you grow soybeans and it gets flooded by 25 feet [7.6 m] of water and you end up with 3 feet [1 m] of oozing mud on top of your pasture, it's going to be a long time before you can plow [plough] – certainly one year. It might be arable [suitable for growing crops] next year, but it would probably be quite difficult to work.'

Overall, economic losses were estimated at US $3–4 billion (about £2–2.5 billion), including delayed river traffic and damage to homes, crops and the fishing industry.

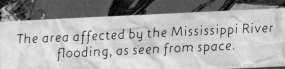

The area affected by the Mississippi River flooding, as seen from space.

REPAIRS AND RECOVERY

Even while the floodwaters were still rising, experts were examining the levees, working out where the system needed repairs to cope with the next flood. For example, they had to repair the levees and fix seepage and sand boils (where water erupts through a bed of sand). It was expected that the repairs would be completed before the next flood season.

Expert view

Make room for the river

'Man must not try to restrict the Mississippi River too much in extreme floods. The river will break any plan which does this. It must have the room it needs.'
Major General Edgar Jadwin, Chief of Engineers who sponsored the plan for Mississippi flood control adopted in 1928.

A 2012 report assessed how well the authorities and communities coped with the flood. It described the great success of the flood-control system but noted areas for improvement, for example, making the levees more resilient (better able to stand up to floods) and carrying out repairs in the areas at greatest risk of floods.

Memphis, Tennessee, the area that suffered the greatest economic damage from the flood.

At 2 a.m. on 7 July 2012, in the Krasnodar region of Russia near the Black Sea, most people were fast asleep. Suddenly, torrential rains began to beat down, causing ferocious flash floods. The equivalent of five months' rain - 275 mm (11 in) - fell in just one night. The town of Krymsk, at the base of a mountain range, was worst affected. Those who woke up dashed into the streets in their nightclothes, grabbing their children and passports. As the water rose, others were trapped in their homes. Some managed to scramble to their rooftops or into trees to escape the flood, but many others drowned in their sleep. The flood surged to an incredible 7 m (23 feet). With vehicles covered in water, survivors were stranded. Pensioner Lidiya Polinina later described her escape: 'We broke the window to climb out. I put my five-year-old grandson on the roof of our submerged car, and then we somehow climbed up into the attic.'

WHAT CAUSED THE DISASTER?

Russia has a high flood risk: it is surrounded by 12 seas and has many rivers. When heavy rains occur, the drainage system cannot cope with the excess water. The 2012 flash flood was mainly caused by a vast amount of rain falling in a very short time. Also, mudflows and mudslides blocked the Adagum River, stopping the flow of water down the river and increasing flooding.

DEVASTATION

At least 172 people lost their lives in the flash floods and 13,000 homes were damaged. The transport system collapsed as floods destroyed traffic lights and mudslides cut off roads and railway lines. Power lines fell and shops closed, so normal daily life ground to a halt. Children at summer camps in the region were left without electricity or supplies; many were evacuated and sent home. The wider economy was hit as ports closed and oil shipments from Novorossiysk stopped. In all, around 30,000 people were affected.

Rainfall?
275 mm (11 in)

How many killed?
at least 172

How many homes damaged?
13,000

How many people displaced?
at least 850,000

Cost to the economy?
US $600 million (£369 million)

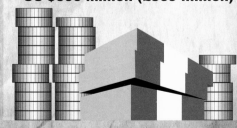

Initial reaction

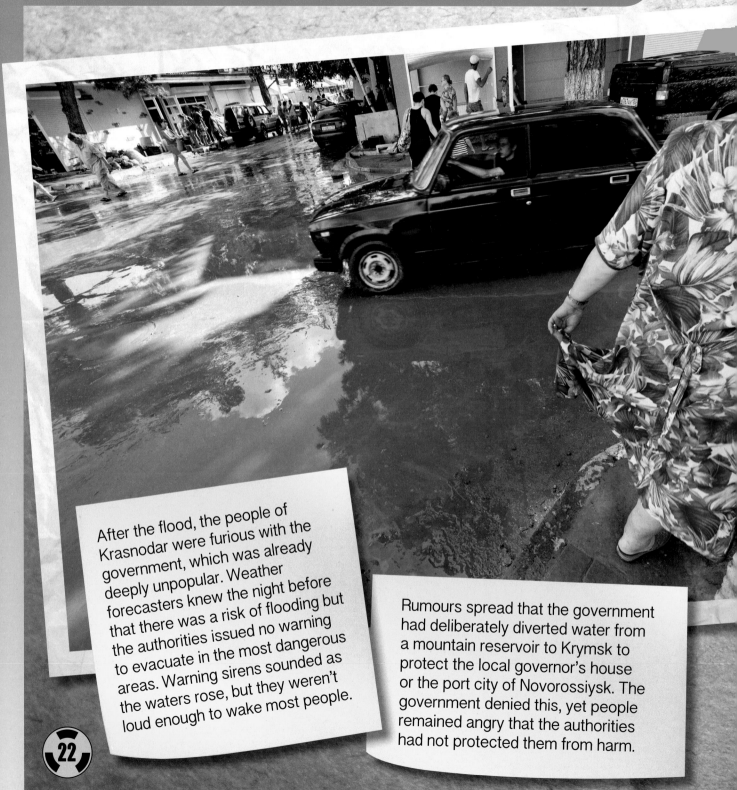

After the flood, the people of Krasnodar were furious with the government, which was already deeply unpopular. Weather forecasters knew the night before that there was a risk of flooding but the authorities issued no warning to evacuate in the most dangerous areas. Warning sirens sounded as the waters rose, but they weren't loud enough to wake most people.

Rumours spread that the government had deliberately diverted water from a mountain reservoir to Krymsk to protect the local governor's house or the port city of Novorossiysk. The government denied this, yet people remained angry that the authorities had not protected them from harm.

The government now took action. President Vladimir Putin visited the area the evening after the flood to view the damage from a helicopter, and the government immediately set up a commission to help the victims and a fund to give money to those who had lost their homes.

Emergency teams were airlifted from Moscow by plane and helicopter. Up to 1,000 rescuers searched for victims and evacuated survivors. A few days later, aid arrived from other parts of Russia and surrounding countries. More than 3,000 Russian volunteers came to help give out supplies. Ordinary people posted information on Twitter about the goods needed and how to send them, and money and donated goods poured into Krymsk. Eight aid distribution centres and a mobile hospital were established. In Krymsk, bath points were set up to allow the newly homeless to keep clean.

Eyewitness

House in ruins

Gelendzhik was flooded after 28 cm (11 in) of rain fell in 16 hours. One man, who returned to his water-sodden home afterwards, explained: 'I climbed the ladder [to the attic]. Otherwise I would have drowned here because you can see that the water was above my height. . . . We don't know what to do. We will have to throw away everything. You can't sleep here, live here, eat anything. I left my house in these clothes, and these are the only things I've got now.'

A flooded street in Gelendzhik.

Learning lessons

Although the government quickly raised money to help flood victims, one month later, many complained that it was hard to access assistance. Some residents said they had to wait hours and fill in many forms, while others said they had not received any money yet. Viktoria Draganova lost her home in the flood and was still in a shelter: 'Autumn is approaching. We keep looking at the sky. It will soon get cold and rainy. What will we do? Where will we go?' Even a year later, some residents were still waiting for compensation.

RECOVERY EFFORTS

However, the government did carry out reconstruction work. Two new neighbourhoods were built in Krymsk, and schools and a daycare centre were rebuilt. Charities continued their relief operations during 2012. The Russian Red Cross distributed aid and assisted local authorities with reconstruction by providing modern equipment for local hospitals, nurseries and schools.

PREVENTION AND PREPARATION

After the flood, work began on expanding and deepening the bed of the Adagum River to double the amount of water it could hold and help to prevent it from overflowing in the future. The riverbanks were reinforced and bridges reconstructed.

Some steps were taken to improve flood preparedness. A Red Cross project started in 2012 to train volunteers in communities in the Krasnodar region to respond to disasters and give assistance to those affected. Yet, in general, Russian communities are not prepared for flooding and the warning systems are poor. There is an urgent need to assess which areas are most at risk and develop methods to protect people.

Local residents watch the river in Krymsk after the devastating flash floods.

Case study

Paying for their mistakes

In 2013, four local officials were convicted of failing to warn residents about the floods, and three were sent to prison. It was stated that the officials 'did not announce the emergency situation in time and did not warn and rescue people.' Afterwards, to cover up their mistake, three of them had created false documents stating that they had issued a warning. The court case showed local authorities how important it is to issue clear flood warnings in future.

Former Krymsk mayor Vladimir Ulanovsky was one of the officials sent to prison.

Prediction and protection

Technology can greatly assist with the prediction of floods. Improvements in weather forecasting and computer modelling make it easier to predict floods accurately. Experts make computer models of flood situations to work out what the effects will be on people and wildlife. Combining the forecast information and computer models, meteorologists work out how likely it is there will be a flood and issue warnings showing the level of risk in particular areas.

FLOOD UPDATES

Communications technologies are used to spread the warnings. During the 2011 Mississippi River flood, officials posted information and updates to the public on Facebook, Twitter, YouTube and Flickr. The US Army Corp Engineer Research and Development Centre in Vicksburg, Mississippi, produced a smartphone app that provided real-time GPS (accurate place and time) information about the progress of the fight against the flood so that flood fighters had up-to-the minute data. The army plan to improve this tool for future use.

LEVEES, DAMS AND BARRIERS

Communities can be protected from floods in various ways. Levees are built alongside a river to prevent it from flowing over its banks and flooding the land alongside. A dam is a solid wall across a river that blocks its flow, storing the extra water in a lake. Barriers can also be useful, such as the Thames Barrier in London. It has gates that close to protect the city from floods.

Expert view

Forecasters can get it wrong

Despite the advances in weather forecasting and computer modelling, experts cannot always predict floods accurately. At the start of September 2013, Colorado weather forecasters knew that rain was on the way. But as Nezette Rydell, head meteorologist at the National Weather Service in Boulder, explained, 'There was a lot of discussion . . . We did not forecast 8 inches [20 cm] of rain.' Extreme weather scientist Kelly Mahoney noted that the computer models failed too. Normally, weather models become more accurate closer to the time of the event. In this case, 'As the event got closer ... things actually started to go off the rails a little bit and show actually less precipitation [rain] forming. The picture got a lot less clear.'

Meteorologists track extreme weather events and do their best to warn people about them.

The storm-water challenge

Technology is also vital for storm-water management. Storm water is rainwater that does not soak into the ground. As it flows, it collects soil, rubbish and dangerous chemicals. This polluted water goes down drains and sewers to the nearest river, harming wildlife.

In the USA, the Green Infrastructure programme uses natural processes to manage storm water. Rainwater harvesting is the collection and storage of rainwater for later use. Downspout disconnection involves using rooftop drainage pipes so that rainwater flows into the garden.

A storm-water management system in Albany, New Zealand

IT'S NOT ALL DOWN TO TECHNOLOGY

As well as prediction and protection technologies, governments, local authorities and individuals need emergency plans for dealing with floods. In 2011, the Mississippi region was well prepared to meet the challenges of the river flood, while in Russia in 2012, the lack of flood preparation caused a terrible loss of life and property.

Floodplain management is crucial too. The growing numbers of people living on floodplains, for example in the USA and Bangladesh, mean more people will be affected when a flood hits. It is best to ensure that activities in areas at high risk from flooding will not suffer when the waters rise, for example, by banning house building on US floodplains or promoting duck or fish farming in Bangladesh.

A catfish farmer in Bangladesh.

Expert view

Don't build on floodplains

In June 2013, southern Alberta, Canada was ravaged by floods. Seven years earlier, former agriculture minister George Groeneveld had called for laws to prevent development in flood-prone areas. This action was not taken, and many homes built since 2006 were flooded. Groeneveld commented, 'People have to stop building on floodplains. If we could have got a handle on that, it would have saved us a lot of grief. There wouldn't have been as many houses taken away in all the towns that have a flooding problem.'

A Calgary resident walks barefoot during the floods of summer 2013.

We know that climate change is making floods more frequent and severe. Our challenge is to develop technologies to improve the prediction of floods and protect communities, and to consider how we use land prone to flooding to reduce the risks of death and devastation.

Glossary

asset Something that is valuable or useful to someone.

authority The people or an organization with the power to make decisions in a country or region.

breached When water goes over riverbanks, it has breached the banks.

climate change Changes in the Earth's temperature, wind patterns and rainfall, especially the increase in the temperature of the atmosphere that is caused by the increase of particular gases, especially carbon dioxide.

compensation Money given to make up for something that has gone wrong.

computer modelling Using a special computer program to model a real-life situation, such as a flood striking the coast.

contaminate To make something dirty by adding a substance that is dangerous or carries disease.

debris All the things that have been swept up by the floodwater, such as wood, household goods and rubbish.

displaced Moved from the usual place.

estuary The wide part of a river where it flows into the sea.

evacuate To move people from a place of danger to a safer place.

export To sell and send goods to another country.

fertility Of soil, the state of being fertile – good for growing crops.

fertilizer A substance added to soil to make plants grow better.

flash flood A sudden flood of water, usually caused by heavy rain.

floodplain An area of flat land beside a river that regularly becomes flooded when there is too much water in the river.

forecast A statement about what will happen in the future, based on information that we have now.

gravity The force that attracts objects in space towards each other, and, on Earth, that pulls them towards the centre of the planet, so things fall to the ground when they are dropped.

insecticide A chemical for killing insects.

landslide A mass of earth, rock or other material that falls down the slope of a mountain or cliff.

latrine A simple toilet with a hole over a pit.

levee A low wall built at the side of a river to prevent it from flooding.

meteorologist A scientist who studies the Earth's atmosphere and its changes, especially for forecasting the weather (saying what it will be like).

monsoon A period of very heavy rain in the summer in South Asian countries such as India and Bangladesh.

mudslide A large amount of mud sliding down a mountain, often destroying buildings and injuring or killing people below.

non-governmental organization (NGO) An organization that is independent of government and business.

polluted When dirty or harmful substances are added to land, air or water so that it is no longer pleasant or safe to use.

prediction A statement that says what you think will happen.

reconstruction Rebuilding something that has been damaged or destroyed.

reservoir A natural or artificial lake where water is stored before it is taken by pipes to houses.

sediment Sand, stones or mud carried by water or wind and left, for example, on the bottom of a river.

storm surge When high winds whip up the water so it rises and floods over the land.

submerge To go under the surface of water or to make something go under the surface of water.

tarpaulin Heavy, waterproof sheet used for shelter.

tropical cyclone A violent storm in which strong winds move in a circle, and which happens in tropical areas – places near the Equator, in the middle of the Earth.

Find out more

Books
Non-fiction
Developing World: Russia and Moscow by Philip Steele (Franklin Watts, 2013)
Eyewitness Disaster: Floods! by Helen Dwyer (Franklin Watts, 2011)
It's Cool to Learn about Countries: Bangladesh by Tamra Orr (Cherry Lake Publishing, 2010)
Natural Disasters: Floods by Chris Oxlade (Wayland, 2010)
Wild Weather: Floods by Angela Royston (QED Publishing, 2009)

Fiction
Flood by Alvaro F. Villa (Capstone, 2013)
Flood Child and *Flood and Fire* both by Emily Diamand (Chicken House, 2009 and 2010 respectively)
Manu's Ark: India's Tale of the Great Flood by Emma Moore (Mandala Publishing, 2012)

Websites
Bangladesh (LEDC)
http://www.coolgeography.co.uk/A-level/AQA/Year%2012/Rivers,%20Floods/Flooding/Bangladesh.htm
Causes and effects of flooding, and some solutions

Flood Pictures: Mississippi River at its Worst
http://news.nationalgeographic.co.uk/news/2011/05/pictures/110510-mississippi-river-memphis-tennessee-flood-record-crests-nation/
Images of great Mississippi River floods past and present

Mississippi River Flooding
http://htekidsnews.com/flooding/
How a levee was broken to protect the town of Cairo from floods

Planet Earth: Floods
http://www.kidcyber.com.au/topics/floods.htm
Basic information with links to the science of floods, flood fighters and survivors

Russia flash floods: 144 killed in Krasnodar region
http://www.bbc.co.uk/news/world-europe-18751198
News report about the floods, 8 July 2012

What is flooding?
http://news.bbc.co.uk/cbbcnews/hi/newsid_1610000/newsid_1613800/1613858.stm
Children's BBC guide to floods

Videos
Floods in Bangladesh
http://www.bbc.co.uk/learningzone/clips/floods-in-bangladesh/8296.html
A report about the annual floods

Landslides and More: Floods
http://video.nationalgeographic.co.uk/video/environment/environment-natural-disasters/landslides-and-more/floods/
How floods occur

'Over 170 killed as tsunami-like flood hits southern Russia'
http://rt.com/news/toll-dead-region-floods-633/
Video and report, with photos

Index

Adagum River 21, 25
aid 11, 13, 23, 24
armies 11, 27
authorities 11, 16, 22, 25, 30

Bangladesh 5, 8–13, 29
barriers 27
  dams 27
  levees 15, 27, 30
buildings
  damage to 6, 9, 15, 18, 21
  rebuilding 13, 18, 24, 29

charities 13, 24
climate change 7, 9, 30
coastal floods 4, 5, 7, 8
communication 23, 27
computer modelling 26, 27, 30
costs 12, 15, 17, 18, 21
crops, damaged 9
cyclones 5, 7, 30

death toll 6, 9, 10, 15, 21
debris 6, 18, 30
deforestation 9
diseases 6, 10
drains 4, 21

emergency services 11, 23
evacuation 4, 15, 21, 23, 30

farming 6, 9, 12, 13, 18, 29
flash floods 4, 5, 8, 15, 20, 30
floodplains 5, 6, 29, 30
food supplies 10, 11, 12
forecasting weather 22, 26, 27, 30

governments 11, 22, 23, 24
GPS information 27

homelessness 9, 15, 21, 23
hospitals 23, 24

ice, melting 5, 7

jobs 12

Krymsk 20, 22, 23, 24, 25

landslides 10, 30
levees 15, 16, 19, 30
lightning 10

Maldives 5
medical supplies 11, 23, 24
meltwater 5, 7, 14
meteorologists 26, 27, 30
Mississippi River 6, 14–19
monsoons 5, 8
mud 6, 11, 18, 21
  mudslides 8, 21, 30

Nile, River 6

pollution 6, 18, 28, 30
power 21
prevention 15, 16, 17, 25, 29
Putin, Vladimir 23

rainfall 7, 8, 9, 15, 20
rainwater harvesting 28
rivers 4, 5, 6, 14–19, 25
Russia 20–25

sand boils 19
sea levels 7
sediment 6, 8, 18, 30
seepage 19
sewage 6, 28
shelters 10, 11, 17
smartphone apps 27
social media 23, 27
South Asia 5
spillways 15, 16
storm surges 5, 30
storm water 28–29
Sylhet 9, 10, 11, 13

technology 15, 26, 27, 28, 29
Tennessee 15, 19
Thames Barrier 27
thunderstorms 5
tides 7
toilets 10
transport 8, 9, 21

USA 6, 14–19, 27, 28

volunteers 11, 23, 25

warning systems 22, 25, 26
water supplies 10, 11
waves 5, 17
wildlife 28
wind 5, 7